# Life in a Tide Pool

William M. Stephens

# Life in a Tide Pool

illustrated by ARVIS STEWART

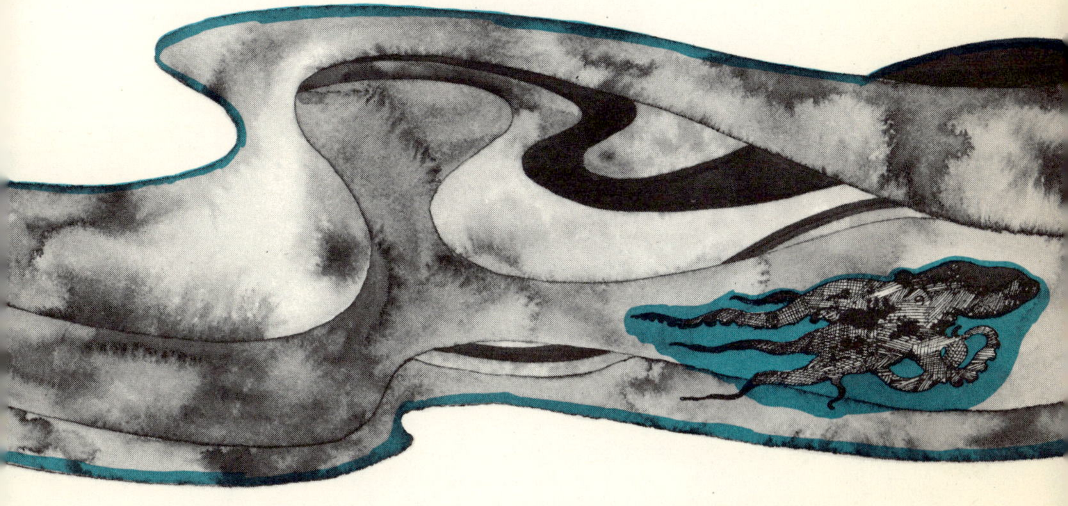

McGraw-Hill Book Company

*New York St. Louis Auckland San Francisco Düsseldorf Johannesburg Kuala Lumpur London Mexico Montreal New Delhi Panama Paris São Paulo Singapore Sydney Tokyo Toronto*

Library of Congress Cataloging in Publication Data

Stephens, William M date
Life in a tide pool

Includes index.
SUMMARY: Explores the struggle for life of the plants and animals in a tide pool.
1. Tide pool ecology—Juvenile literature.
[1. Tide pool ecology. 2. Ecology] I. Stewart, Arvis L., illus. II. Title.
QH541.5.S35S78 574.92 74-31173
ISBN 0-07-061186-6 (lib. bdg.)



1234567 RABP 7898765

# Contents

*For the children of Pimplegaon Village, Maharashtra State, India, with love.*

# *Midday at the Tide Pool*

It is midday and the tide is out. The sun is hot; the temperature of the water in the tide pool is nearly 100 degrees F. The small pool—only about ten feet long and five feet across—feels like bath water, and some of the creatures living in the pool are in distress.

Popeyed gobies and blennies lie still on the bottom, their mouths open and gill covers flared. They are pumping water rapidly over their gills in an effort to get more oxygen. Some of the other fishes are swimming slowly at the surface of the water, sometimes even taking a gulp of air.

The oxygen level of the water is very low: much of it has been used up by the fishes, crabs, snails, and other animals that live in the pool. The last high tide was before dawn. For about nine or ten hours the pool has been cut off from the sea.

# The Tide Rises

But the tide is rising. Gradually it creeps up the pitted rock shelf. Already some of the other pools, closer to the sea, have been flooded. Finally a wave surges up and pours gallons of cool water into the pool. Soon another wave reaches the pool. The water pours over the rocky lip of the pool's edge like a miniature green waterfall, churning the surface and filling the water with oxygen-rich bubbles of air.

The animals of the pool respond to the fresh sea water and become active. An octopus jet-propels itself out of its cave to capture a crab, which it drags back to its lair. The gobies and blennies no longer labor in their breathing. One three-inch blenny has

The fish at the lower right is a blenny and behind it is a goby. A sea anemone grows on a shell, and in the distance are the arms of an octopus.

hopped up to the sloping rocky shelf to face into each wave as it surges into the pool. A two-inch goby is now on the vertical rock wall, held there by the suction disk on its belly, its bulging eyes turned toward the surface. From time to time the blennies and gobies dart forward and snap their jaws. They are feeding on tiny transparent shrimps and other little creatures that drift into the pool with the moving waters. These creatures are called plankton.

Now a big red hermit crab rouses himself and stands up, throwing off his heavy conch-shell dwelling. Atop the shell is a large sea anemone. The anemone is resting now. Its tentacles are withdrawn and it looks like a little brown bag closed tightly by a drawstring. Soon the anemone opens up to feed. It unfolds, like the petals of a flower, and its long orange tentacles stretch out. As the inpouring water sloshes back and forth in the pool, the tentacles wave like pine needles in a breeze.

The fresh sea water is alive with plankton. Some of the little drifting animals are eaten by the anemone, which traps the plankton with its sticky tentacles. The tentacles are covered by many stinging cells called nematocysts (Ne-mat-o-sists), which kill or paralyze the prey.

All of the bivalve mollusks in the pool—oysters, clams, mussels—have also started feeding. During most of the day—while the pool has been cut off from the sea—they have kept their shells closed; but now, with a fresh supply of plankton-filled sea water, they are busily at work pumping the water through their bodies and straining out tiny organisms for food. The oysters and mussels are attached to the sides of the pool; the mussels lower down, where they are always submerged by water, and the oysters higher, because they can tolerate being out of water for short periods.

There are several types of plant plankton in the circle. The others in the rectangle are animal plankton.

# The Barnacles

Near the oysters on the rocks are many thousands of acorn barnacles. In some places there are more than a thousand barnacles in a square foot of space. Each barnacle is like a tiny volcano—a cone-shaped mound of limestone. Some of the mounds gape open and are hollow inside, for the barnacle animal is no longer inside the shell. Something has eaten it: an oyster drill, perhaps, a whelk, or some other predator. But the outer shell, firmly cemented to the rock, still remains. And inside these round stone houses live tiny creatures. Inside one, a slender yellow fish is curled up; in another lives a purple sea urchin. Others contain worms, crabs, and shrimps.

Now the living barnacles are feeding. At the top of each mound they have opened their sliding doors and extended their feathery feet, or cirri, which they rake through the water to catch plankton. When the cirri come into contact with tiny creatures, the prey are swept into the barnacle's mouth. The actual barnacle animal is much like a small shrimp, but it cannot leave its hard-shell home.

A close-up of barnacles. The one on the far left has a sea worm in its empty shell.

# The Flooded Pool

By dusk, the tide is nearly full and the pool is flooded. For an hour or so it is not a pool at all; it is the ocean. Fishes from the open sea now cruise about the pool, probing here and there for food. For a few minutes a school of silver herring fills the water, like a thousand slivers of glass. They are feeding on microscopic plankton. Several pompanos flash by, followed by a slender two-foot barracuda. A chocolate-brown nurse shark swims slowly through the pool, pausing to crush and eat a mussel before disappearing over the edge of the pool. Colorful wrasses, damselfish, and butterfly fish flutter about;

A sea anemone is on an empty conch shell, and above and to the right is a pompano. To the lower left is a damsel fish, in the very upper left are two wrasses, and the thin fish at the upper center is a barracuda. Below it are two scaly parrot fish. The fish with the black circle on it is a butterfly fish, and to its left is another damsel fish.

two red parrot fish eat green plants and occasionally bite off pieces of the rock wall itself.

As the water level begins to fall, most of the visitors move on. Sometimes a visitor becomes trapped in the pool and takes up residence there, remaining even at high tide, when it could easily leave. More often a trapped visitor from the open sea fails to survive in the pool and soon dies, to furnish food for the gobies and blennies and other creatures.

Few sea animals are hardy enough to live in a tide pool. They must be able to tolerate extreme conditions. The temperature of the small body of water changes rapidly. The saltiness, or salinity, also changes. When the hot sun evaporates much of the fresh water, the remaining water becomes very salty, and this may kill some animals. A hard downpour of rain, on the other hand, can make the pool too fresh for some sea creatures.

# The Tide is Out

Soon the pool is separated from the sea. The water in the pool is at its deepest—about two feet in some places—and is full of plankton. The animals of the plankton are varied, but now the most abundant are the barnacle larvae. These larvae have been floating and drifting in the sea for weeks, passing through six different stages. Now they are in the cypris, or settling, stage.

Each cypris has a bivalve shell and twelve legs. After drifting for hundreds of miles in the open sea, these barnacle larvae are ready to settle down and stay in one place for the rest of their lives. They have reached the stage in their life when they must find a suitable surface to cement themselves to or they will die.

Barnacle Larvae about to settle down for life.

One of the larvae drifts through the forest of colorful anemone tentacles that wave atop the shell of the big red hermit crab and barely misses becoming food for a mussel as he settles down among a cluster of acorn barnacles. Now he begins to crawl about, moving by means of his two antennae. He will swim no more. With his antennae he probes the surface of the rock and something tells him to stay right there.

Somehow young barnacles sense when they have found a suitable place to settle. Apparently they detect a substance released by barnacles that have settled there before. If the area a cypris tests first does not seem right, it creeps along, seeking another place. The choice of a place is very important, both to its survival and to the survival of its species. A barnacle must be close enough to other barnacles to be able to mate with them, and once fixed to a rock, it can never move from that spot.

Most barnacles are hermaphrodites (Her-maf-ro-dites), containing both male and female parts. But usually they do not fertilize themselves. They fertilize a neighbor by means of a long slender tube that stretches between them—sometimes extending several inches.

A sea urchin, and at the left is a dog whelk. On the rocks to the right are seaweed leaves, barnacles and an oyster drill at the top of the rock.

The surface of the rock is occupied by barnacles of all sizes and by green fuzzy plants—algae—that even grow on the outside of the barnacle shells. But here and there on the rock surface are little furrows, like trails, where a sea urchin has recently fed. Many sea urchins feed every day on the green scum of algae that grows in all sunny parts of the pool.

# Night Time

The tide has gone down and night has come. The water is quite calm in the pool, and the barnacle larva has no trouble holding onto the smooth rock at the place he has selected. He clings to the rock with his twelve legs and two antennae. He uses special glands to produce a brownish cement that seeps through sucking disks on his antennae, binding the antennae firmly to the rock. Once the cement has hardened, even the strongest storm waves will not dislodge him.

The cypris larva does not look at all like an adult acorn barnacle. He does not yet have the house that is shaped like a volcano. But now that he is fastened to the rock, he will proceed to metamorphose (Met-a-mor-fose: change form) in order to become an entirely different-looking creature.

During this process the young barnacle flops onto his back and becomes permanently fixed in that position. He loses his juvenile form and flattens into a shapeless glob. Then, within a few days, he encases himself in a cone of overlapping plates—a smaller version of the volcano shape of the older barnacles.

But this particular cypris larva will never build his protective cone. It is early morning, and the cypris has been attached to the rock for only a few hours. Now a grazing sea urchin, plowing a path through the fuzzy green carpet of algae, scrapes the young barnacle off the rock and swallows him.

# The Sea Urchins

The sea urchin is a vegetarian and not a barnacle eater. But while scraping algae off the rocks, he often consumes tiny barnacles, snails, or worms. Climbing the vertical rock wall to the edge of the algae coating, the sea urchin works back and forth. He feeds with his five sharp-pointed teeth, which fit snugly together in the mouth—the round opening in the center of his underside. He moves by means of tube feet. These are slender, flexible tubes with a suction disk at the end of each one. The urchin operates dozens of tube feet at once as he glides smoothly over the steep walls of the pool's edge.

The urchin's spines cover the outside of his body, so that he looks something like an old-fashioned pincushion. The spines are for defense only. They protect the fragile shell that surrounds the soft parts of the animal. The shell is of calcium carbonate (lime) but is quite thin and can be easily crushed.

Now the sea urchin is half out of water as he grazes on the algae growing near the high-tide mark. The fuzzy green mat of algae is thicker near the surface, where the sunlight is strongest. The urchin is all alone here. Most of the other urchins are on the bottom or hidden in the crevices and holes. Through eons of time a certain caution has been bred into the species. But as with all creatures, some individuals are bolder—or more reckless—than others.

This urchin is too bold to survive. He attracts the attention of a sea bird—a gull—that is flying above the tide pools looking for food. The gull pauses in flight, peering below, then banks and swoops downward. Hovering just above the surface of the pool, flapping his wings rapidly, he darts his head out and grabs the urchin. With his large, strong beak he pulls the urchin off the rock wall easily; yet he does not bite down hard enough to be hurt by the sharp spines.

Holding the urchin carefully in his mouth, the gull soars up above the rocky ledge. Then he opens his bill and lets the urchin fall. Dropping the urchin is not an accident. It is done purposely. For when the urchin strikes the rocks below, the thin shell is crushed and a mass of bright orange eggs spills out. The gull lands and eats the soft, tasty eggs. He also eats some of the other soft parts of the urchin's body. Then, calling shrilly, the gull flies away.

The broken shell of the urchin lies on the rock by the pool's edge. The high tide is now flooding the pool again. Soon a wave washes the crushed shell into the pool, and tiny blennies, gobies, striped sergeant majors, and yellow-and-black angelfish dart about in excitement as they feed on the remains of the urchin.

Joining in the feast is a little brown triggerfish that has just entered the pool in a floating clump of seaweed. The triggerfish is new to tide-pool life. Until the wind blew the clump of seaweed to the shore, he had known only the seaweed jungle, where he caught plenty of tiny shrimps and crabs. Now he is in a strange world. He has never seen creatures like those that live in the pool. He feeds recklessly, as triggerfish do, and pays no attention to the waving tentacles of the big orange sea anemone.

# The Anemone

The anemone is still swaying back and forth on the top of the conch shell that serves as a home for the red hermit crab. The crab has been attracted by the excitement in the pool and comes to see what food can be found in the dead urchin's shell.

The triggerfish has never seen an anemone before and has no reason to fear it. When he brushes against the waving tentacles, he is quickly trapped. The sticky tentacles hold the fish fast, and thousands of tiny darts (nematocysts) squirt poison into his body. The fish is paralyzed and cannot move

as the anemone draws him into his pouchlike stomach.

Inside the anemone's stomach the fish is digested and partially eaten. After a time, the remains are cast out by the anemone and are eaten by the red hermit crab, who is waiting.

The close association of the sea anemone and the hermit crab is beneficial to both creatures. In a way, they are partners. The anemone, by remaining fastened to the hermit crab's "house," gets a free ride to all the places the crab goes. In return, the crab is protected, since the anemone might sting a creature that tries to attack the crab. Also, as we have seen, the two animals sometimes share a meal.

# *Two Boys Visit the Tide Pool*

Neither the crab nor the anemone has much protection against humans who collect things in tide pools. And at the next low tide, that afternoon, two boys wade into the pool, overturning rocks and creating chaos among the life of the pool. When the boys start splashing in the water, all the animals try to hide. The octopus squeezes itself back into the deepest tunnel, and the gobies dart under the ledges. The anemone pulls itself inside its pouch, so that it looks like a large brown bump on the conch shell. The hermit crab withdraws as far as possible into the conch shell.

To the right is an octopus, a parrot fish is in the middle, and a lady crab is over a dying crab. Beneath them is a shrimp.

"Look, a big shell!" one of the boys shouts. "It's a queen conch!"

"Let me see," says the other. "Hey, there's an animal inside!"

"Yeah. Look at the size of that claw!"

They shake the shell, pound it on the rocks, and try to pry the crab out with a stick. Finally they use a large rock to knock a hole in the shell. But still the hermit crab will not come out. After a while, the boys give up in disgust. Tossing the broken shell back into the pool, they go off looking for excitement somewhere else.

The red hermit crab is badly hurt. His soft abdomen was injured by the pounding. Dying, he pulls himself from his shell. Instantly he is set upon by little fishes, shrimps, and crabs, which attack the crab's soft abdomen. This is nature's way. An injured animal is fair prey for all. Unfit to survive, it furnishes food for others.

Soon the octopus reappears, jetting itself across the pool and settling atop the dead crab's body. Flashing red and brown and white in excitement, the octopus claims the carcass and drags it back to his lair. He eats part of the hermit crab and saves the rest for later.

The other creatures now scatter to various parts of the pool. One small goby that had shared in the feast darts into the empty conch shell. The goby has found a new home. From time to time he goes out to feed, but always returns to the conch shell. While going in and out, he is careful to avoid the long, waving tentacles of the anemone, which is still fastened to the shell.

The tiny goby has found a new home.

Goby moves over dry land to get back to his pool.

# The Goby

On the high tide the next morning the ocean again floods the pool. The goby goes out to search for food and swims some distance away. When the tide starts falling, the little fish is trapped in another pool. To get back to his home, the goby jumps out of the strange pool onto dry land. Hopping and squirming his way over the rocks, he makes his way back to his home pool.

The goby is one of only a very few kinds of fish that can actually travel over land for short distances. Scientists do not know how the little gobies can find their way home when they travel over land. Perhaps they follow certain "paths" across the rocks—paths they have selected while swimming about at high tide.

Reaching his own pool, the goby plops into the water, then swims straight to his conch shell home. Darting inside the shell, he is instantly devoured by a spotted moray eel that had come into the pool on the high tide and claimed the empty conch shell for its home.

The moray eel is still hungry and he soon goes out to look for more food. Morays are strong, aggressive fish that have very sharp teeth. They often stay in dark tunnels and holes—the same kind of places where octopuses live.

# The Octopus and Moray Eel

Octopuses and morays are natural enemies. Each one has a definite territory he claims as his own. And each one is ready to fight to the death when his territory is invaded. So when a moray eel and an octopus find themselves together in a small tide pool, there is likely to be trouble.

As soon as the moray leaves the conch shell and goes on the prowl for food, the octopus gets ready for battle. First he stuffs the remains of the dead crab into the back of his lair. Then he moves to the doorway of his cave. He can smell the moray eel, which is

Octopus and moray eel fight to the death

coming his way, attracted by the odor of the dead crab.

Snakelike, the eel slithers into the tunnel. Quickly the octopus is upon him, winding his eight arms around the eel's head, feeling for the eel's eyes with his rasping suction cups. The moray bites and twists, tying his own body into knots, then pulling himself backward through the loops in an effort to dislodge the octopus. As they writhe and fight, the water turns dark with octopus "ink" (a kind of smoke screen produced by octopuses under stress). In the whirling melee, the eel rips and tears the octopus with his teeth while the octopus bites the eel repeatedly. An octopus's mouth is tiny but its beak is razor sharp, and it has a poisonous bite. Slowly the poison takes effect on the eel, and it becomes weaker and weaker. Finally the fish stops struggling and lies still on the floor of the pool.

The octopus is the victor, but at a terrible price. Two of his arms have been torn off and another is torn to shreds. The octopus has turned white as snow, and he breathes very hard, pumping water rapidly in and out of his funnel. He moves very slowly along the bottom of the pool, as if unsure of what to do. From time to time he rolls up in a ball, while waves of color—pink and green and yellow—pass over his mutilated body. He is barely alive when the next high tide sweeps a cascade of green water into the pool. The surge of the water moving back and forth in the pool carries the octopus with it. He seems to have no strength to swim.

# The Barracuda

Finally a wave sweeps the helpless octopus from the pool and carries him out to sea. Quickly a cruising barracuda seizes the limp form and swallows it.

The barracuda is sleek and beautiful. About five feet in length, he has sharp teeth and strong jaws. With one bite he can cut a large fish cleanly in two. He is known to fishermen as the tiger of the sea.

As he moves into deeper water, the barracuda is attracted by a flashing object ahead. It is so bright and interesting that he does not notice the brass swivel and long line curving ahead of it. Neither does he associate the bright object with the droning sound of an outboard motor.

The flashing object is a type of fishing lure known as a spoon. A large hook is attached to the spoon, which is being trolled behind a fishing boat.

The barracuda has never known restraint. He has never attacked any fish he could not

A barracuda attacks a dying octopus.

handle with one bite of his razor-sharp jaws. He has no way of knowing that his life in the sea is about to come to an end.

The bars and spots on the barracuda's body are now bold and distinct. With a sweep of his broad tail he surges forward and opens his great jaws to engulf the bright flashing object.

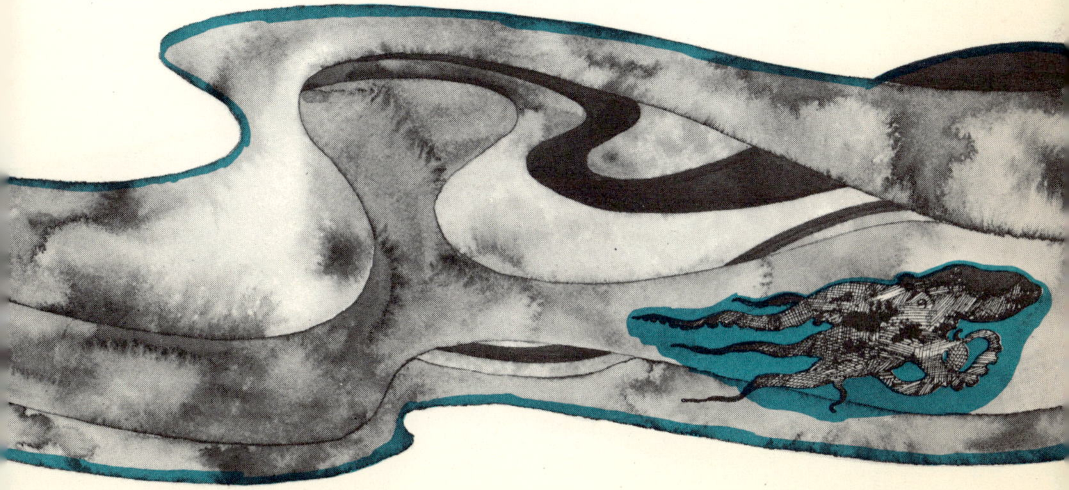

# Index

## *ABOUT THE AUTHOR*

***William M. Stephens*** is a lawyer and naturalist, and the author of many books about nature science and the sea. He is a well-known photographer and scuba diver.

A native of Chattanooga, Tennessee, he received his Doctor of Jurisprudence degree from the University of Tennesee, and his Master of Law degree (in Ocean and Coastal Law) from the University of Miami. He has also studied and worked in oceanography and has participated in research cruises to the Sargasso Sea, the Bahamas, and the Caribbean.

Currently, Dr. Stephens is a Professor of Law at Gaston College in North Carolina and is Chairman of the Prison Reform Committee of the Concerned Citizens for Justice.

## *ABOUT THE ARTIST*

***Arvis Stewart*** was born in Turkey, Texas, and received his art training at Texas Technological College in Lubbock. Mr. Stuart has worked in design studios in New York City and is currently a full-time illustrator. Besides illustrating numerous books, and magazines and other materials, he is involved in painting and print making.

Mr. Stuart, his wife and two children live in Port Jervis, New York.